I0470691

Special Report:
Multiple Aid: Lessons Learned
from the California System

Report by: Carolyn Perroni

This is Report 042 of the Major Fires Investigation Project conducted by TriData Corporation under contract EMW-90-C-3338 to the United States Fire Administration, Federal Emergency Management Agency.

Homeland Security

Department of Homeland Security
United States Fire Administration
National Fire Data Center

U.S. Fire Administration Fire Investigations Program

The U.S. Fire Administration develops reports on selected major fires throughout the country. The fires usually involve multiple deaths or a large loss of property. But the primary criterion for deciding to do a report is whether it will result in significant "lessons learned." In some cases these lessons bring to light new knowledge about fire--the effect of building construction or contents, human behavior in fire, etc. In other cases, the lessons are not new but are serious enough to highlight once again, with yet another fire tragedy report. In some cases, special reports are developed to discuss events, drills, or new technologies which are of interest to the fire service.

The reports are sent to fire magazines and are distributed at National and Regional fire meetings. The International Association of Fire Chiefs assists the USFA in disseminating the findings throughout the fire service. On a continuing basis the reports are available on request from the USFA; announcements of their availability are published widely in fire journals and newsletters.

This body of work provides detailed information on the nature of the fire problem for policymakers who must decide on allocations of resources between fire and other pressing problems, and within the fire service to improve codes and code enforcement, training, public fire education, building technology, and other related areas.

The Fire Administration, which has no regulatory authority, sends an experienced fire investigator into a community after a major incident only after having conferred with the local fire authorities to insure that the assistance and presence of the USFA would be supportive and would in no way interfere with any review of the incident they are themselves conducting. The intent is not to arrive during the event or even immediately after, but rather after the dust settles, so that a complete and objective review of all the important aspects of the incident can be made. Local authorities review the USFA's report while it is in draft. The USFA investigator or team is available to local authorities should they wish to request technical assistance for their own investigation.

For additional copies of this report write to the U.S. Fire Administration, 16825 South Seton Avenue, Emmitsburg, Maryland 21727. The report is available on the Administration's Web site at http://www.usfa.dhs.gov/

U.S. Fire Administration

Mission Statement

As an entity of the Department of Homeland Security, the mission of the USFA is to reduce life and economic losses due to fire and related emergencies, through leadership, advocacy, coordination, and support. We serve the Nation independently, in coordination with other Federal agencies, and in partnership with fire protection and emergency service communities. With a commitment to excellence, we provide public education, training, technology, and data initiatives.

ACKNOWLEDGEMENTS

Interview with the following individuals contributed important information and perspective for this report.

Richard Adams	Assistant Director for Fire Operations, U.S. Forest Service – Washington, D.C.
C. Richard Aronson	Chief State Fire and Rescue Coordinator, California Office of Emergency Services
Stanley C. Lake	Deputy Chief for Cooperative Fire Services, California Department of Forestry and Fire Protection
Lloyd M. Limprecht,	Division Chief for Fire Protection Planning, California Department of Forestry and Fire Protection
Dave Mack	Chief of Fire Prevention, California Department of Forestry and Fire Protection
William M. Medegovich	Regional Director, Federal Emergency Management Agency – San Francisco
William J. Patterson	Program Specialist, Natural and Technological Hazards Division, Federal Emergency Management Agency – San Francisco
Donald G. Perry	Deputy Chief, Santa Barbara County Fire Department
Richard Peterson	Chief, Santa Barbara County Fire Department
Mike Sherr	Deputy Chief State Fire and Rescue Coordinator, California Office of Emergency Services
Keith E. Simmons	Operations Chief, Santa Barbara County Fire Department
Robert L. Solari	Regional Mobilization Coordinator, Aviation and Fire Management, US. Forest Service – San Francisco
Kenneth J. Stanley	Deputy Chief for Fire Control Operations, California Department of Forestry and Fire Protection
Dick Starr,	Deputy Chief State Fire and Rescue Coordinator, California Office of Emergency Services
William C. Teie	Deputy Director for Fire Protection, California Department of Forestry and Fire Protection
Jack W. Wiest	Staff Chief for Fire Planning, Research and Cooperative Fire, California Department of Forestry and Fire Protection

TABLE OF CONTENTS

INTRODUCTION

Mutual aid – help among neighbors – is an integral part of emergency response. Mutual aid becomes more important and more complicated as the magnitude of emergency incidents increases and the size of individual community budgets decreases. Given the current economic and social climate, it is simply unrealistic to assume that a single community has all the resources required to cope with any and all emergencies it may face.

The State of California has developed over the last 40 years an exemplary Statewide Fire and Rescue Mutual Aid System. Designed as part of the State's overall multi-hazard emergency response, the mutual aid system has been used in a wide range of fire and non-fire incidents.

The system is not unique; many States have mutual aid systems. However, because it is exercised continually on the plethora of wildland fire incidents in the State, the system has achieved a degree of sophistication not often seen in other systems.

A testament to the respect the Fire and Rescue Mutual Aid System has gained within the State, similar systems have been designed, or are being developed, for police, emergency medical services, public works, and other disciplines in the State that have a role in emergency planning and response.

This report includes California's Fire and Rescue Mutual Aid System and its evolution, how mutual aid is used in major fire and non-fire incidents, and discusses some reasons for its success over the years. The purpose of the report is to share lessons learned to help other States advance their systems and give the fire service in general some useful ideas on providing mutual aid.

HOW THE MUTUAL AID SYSTEM WAS DESIGNED

History

The first *California Fire Service and Rescue Emergency Mutual Aid Plan* was prepared and adopted in 1950 as an annex of the *California State Civil Defense and Disaster Relief Plan*. The plan, and the mutual aid system which evolved from it, derive their authority from the "California Emergency Services Act of 1970" (which superseded an earlier "California Disaster Act"), the California Master Mutual Aid Agreement, Section 3211.92 of the State of California Labor Code, and Section 8690.6 of the State of California Government Code.

The concept of mutual aid in California is based on a common understanding that every individual community cannot gear up for every possible emergency, and that the best way for communities to make sure help is available when they need it is to be available to give help when others need it. The system creates a structural framework for offering and receiving assistance when a community's emergency response needs outstrip its own available resources.

Original development of the mutual aid system was funded with Federal grant money. In 1951, the Federal Government established a program to match State and local funds for the purchase of fire and rescue apparatus and equipment. It was under this program that California purchased just over 100 pieces of apparatus, including 1,000-gpm fire engines and 29 heavy rescue vehicles.

California Office of Emergency Services (OES) then allocated this apparatus among individual fire departments strategically placed throughout the State. The departments could use the apparatus (1) to respond to multiple-alarm fires within their own boundaries; (2) to temporarily replace any of their own first-line fire apparatus out of service due to mechanical problems; or (3) to loan to other departments in their operational area to replace equipment out of service for repairs. The only condition for receipt of the apparatus was the department's agreement to use it when called upon to render aid through the Fire and Rescue Mutual Aid System.

Over the years reliance on the OES apparatus has changed, and there is no question that the system would continue to work even if the apparatus were no longer available. However, the initial purchase and allocation of the apparatus has been credited with providing an impetus for participation in the Statewide mutual aid system.

The Master Agreement

The California Disaster and Civil Defense Master Mutual Aid Agreement, developed along with the original State disaster plan, serves as the mechanism for implementing the *California Fire Service and Rescue Emergency Mutual Aid Plan*. (See Appendix A.) All 58 counties and nearly all local governments within the State are signatories to the agreement. These signatories represent more than 1,200 fire service organizations and include every segment of public fire protection.

The master agreement covers the loan of equipment, facilities, and personnel in a variety of emergency situations, not only those resulting from fires. It provides for mutual aid "to prevent and combat the effect of disasters which may result from such calamities as flood, fire, earthquake, pestilence, war, sabotage, and riot," including assistance in the areas of "rescue, relief, evacuation, rehabilitation, and reconstruction."

Recognizing that disasters are unpredictable and may strike anytime, anywhere, the crafters of the agreement were careful to stipulate that "no party shall be required to deplete unreasonably its own resources, facilities, and services in furnishing mutual aid." This is important. Most local mutual aid calls take units out of their home departments for only a few hours, and making adjustments to compensate for their departure is fairly routine. However, resources contributed to respond through the Fire and Rescue Mutual Aid System could be away for extended periods on major incidents, making the back-filling operation more complicated.

The agreement also reinforces the position of local officials during incidents in their jurisdictions. Under the agreement, the responsible local official in the jurisdiction requiring mutual aid assistance remains in command of the incident response and directs the use of personnel and other resources provided through the Fire and Rescue Mutual Aid System.

The agreement stipulates that each signatory will:

(1) develop a plan for effectively mobilizing its resources and facilities, both public and private, to cope with any type of disaster and to submit it to the State Disaster Council for approval;

(2) provide mutual aid assistance through the system without reimbursement unless otherwise provided for in existing agreements between parties, legislation, or regulations;

(3) furnish resources and facilities in emergencies – fire, police, medical and health, communications and transportation – to each other in accordance with duly adopted mutual aid plans as

long as furnishing such aid does not unreasonably deplete its own resources, facilities, and services;

(4) conduct raining and exercises in accordance with duly adopted mutual aid plans to ensure that it can provide effective assistance when called upon;

(5) furnish mutual aid "in all cases of local peril or emergency and in all cases in which a STATE OF EXTREME EMERGENCY has been proclaimed"; and,

(6) abide by mutual aid agreements the State of California enters into with other States or the Federal Government.

To facilitate effective and efficient response, an inventory of all fire service personnel, apparatus, and equipment in the State is maintained and updated annually by the California Office of Emergency Services (OES). The inventory is published (it is currently being computerized), and OES provides it to operational area (local) and Regional coordinators.

Over the years, the Fire and Rescue Mutual Aid System has been reviewed and revised by the California OES Fire and Rescue Service Advisory Board/FIRESCOPE Board of Directors to keep pace with changing emergency response needs in the State. Today, many private fire protection organizations participate in the system as well.

Although the system is exercised most on fire-related incidents, particularly wildland fires, it was designed, and has proven, to be useable for multi-hazard response.

The Structure

The structure of the California Fire and Rescue Mutual Aid System builds on the voluntary, community-to-community mutual aid common throughout the history of the fire service. California has taken this basic, traditional concept of mutual aid one step further by adopting "automatic aid." Automatic aid means that, on a predetermined basis, first alarm response is provided by the nearest fire station, regardless of community boundaries. As one fire service official put it, "the public doesn't care what color the engine is or where it comes from; they just want to be assured of getting the fastest response when they're in trouble."

Fire departments in most parts of California have effectively used mutual aid on a day-to-day basis for many years. The systems and attitudes that have structured the Regional and Statewide systems have also supported the regular use of mutual aid among departments.

Mutual aid was originally based on the concept of fire departments assisting each other with incidents that are beyond the capability of local resources. In California, in particular, this concept has often been broadened to include providing efficient protection and prompt response to emergency incidents, without regard to jurisdictional boundaries.

Many of the urban and suburban areas of California are made up of hundreds of incorporated local jurisdictions and unincorporated areas. The complex boundary lines present a very real problem for fire departments that are responsible for protecting irregularly shaped areas, which often include isolated "islands" and difficult access problems. In addition, small jurisdictions often have the responsibility to provide protection for high risk areas, particularly wildland interface areas.

The fire organizations protecting these areas range from small departments, with one or two fire stations, to county and city departments with over 100 stations. Some incorporated local jurisdictions contract with the county fire departments that protect unincorporated areas, while other municipalities operate their own independent fire departments. Virtually all of the departments participate in mutual aid and automatic response programs, based on two-party agreements and systems involving several adjacent jurisdictions.

In several cases local jurisdictions have combined their fire communications centers under joint powers agreements, to share costs and maximize efficiency with their neighbors. Under this system, all emergency calls originating within the multi-jurisdictional area are dispatched from the same center and the closest units respond, without regard to boundary lines.

The communications systems that have been established to support these systems are usually referred to as "Nets," and all of the fire resources within a Net operate as if they were one large fire department. For example, Orange County NET 6 includes the Huntington Beach, Westminster and Fountain Valley Fire Departments. The Huntington Beach Communications Center dispatches all calls for these three jurisdictions, utilizing a computer aided dispatch system and an 800 MHz trunked radio system.

Under the Statewide plans for large scale incidents, strike teams and other resources are assembled from the jurisdictions within a Net in accordance with agreements among the participating jurisdictions. The local jurisdictions within a Net may also agree to share the cost of other services and facilities, such as training centers and maintenance services.

While joint communications centers can improve the efficiency of mutual aid systems and speed the response of units from neighboring jurisdictions, they are not essential to provide effective automatic response and mutual aid systems. Most fire departments in California have established standard operating procedures (SOPs) with their neighbors to support different levels of automatic response and mutual aid assistance on a routine level. When a situation exceeds the resources that are available within the immediate area, the existing system of "operational areas" and "mutual aid regions" is activated for immediate escalation of requests for additional assistance.

The cities of Santa Monica, Culver City, and Beverly Hills are an example of this approach. Each of the three cities provides its own fire communications system and responds to most one-alarm incidents with its own resources. If additional assistance is required, the departments call upon each other and upon the Los Angeles City Fire Department, which is responsible for areas bordering on all three cities.

If an incident exceeds the capabilities that are immediately available from those departments, the request for assistance is directed to the "operational area" which is managed by the Los Angeles County Communications Center. This center can activate resources from the entire county, from the region or from the entire State of California.

Community-to-community mutual aid in California has evolved to include the private sector as well. This includes having protocols for accessing heavy equipment, like bulldozers and other specialized equipment that exists in the private sector, to augment the government-held inventory. In fact, some large companies are like microcosms of communities; they have their own fire and police forces and even their own public works capability. In many cases these companies have developed or acquired state-of-the-art equipment that outstrips what local jurisdictions, and even the State, have.

To meet resource needs as incidents escalate beyond neighbor-to-neighbor mutual aid capability, the State is divided into 65 operational areas (generally along county boundaries) and six mutual aid regions. These geographic divisions are the backbone of the Fire and Rescue Mutual Aid System. (See Appendix B.)

There is a fire and rescue coordinator for each operational area elected by the fire chiefs in that area. A Regional fire and rescue coordinator is elected from among the operational area coordinators in the region. The chief of the OES Fire and Rescue Division serves as the State fire and rescue coordinator, overseeing the system.

Operational area and Regional coordinators volunteer their time. These jobs are performed in addition to the regular jobs they are paid to do in their respective jurisdictions. On a day-to-day basis, they may have to spend relatively little time, but when something happens – a major fire, an earthquake, flood, mudslide – they get very busy.

They serve as the focal point for coordinating and dispatching resources requested through the Fire and Rescue Mutual Aid System during emergencies in their areas/regions. In non-emergency times, they lead efforts to promote development of consistent local fire and rescue emergency plans and compatible communications networks to facilitate operations in emergency situations; maintain an up-to-date inventory of resources within their areas/regions; and coordinate periodic reviews of operations and procedures for lessons learned and needs for improvement.

In most cases, mutual aid in the State of California is provided voluntarily. According to the "California Emergency Services Act of 1970," however, mutual aid is obligatory in a "State of War Emergency" and may be obligatory during a declared "State of Emergency."

Mutual aid also is rendered without reimbursement. Each community takes responsibility for expenses incurred when it helps its neighbors. Given the increasing cost of training and equipping fire departments these days, this shows an enormous commitment on the part of the California fire service to the concept of mutual aid.

There is one exception to gratis mutual aid. A written agreement, commonly called the "five-party agreement," provides for reimbursement for mutual aid that extends more than 12 hours in major incidents on State- or Federal-responsibility lands. Signatories to the agreement are California OES (representing all local governments), California Department of Forestry and Fire Protection (CDF), the U.S. Forest Service (USFS), the U.S. Bureau of Land Management, and the National Park Service. (See Appendix C.)

Incidents covered by the agreement tend to be major wildland and forest fires requiring long-term assistance and may involve several jurisdictions. Areas included in the agreement usually are State and National parks and forests, but it also covers other Federally protected lands and State wildland areas that provide the State's watershed and are not covered by an incorporated city.

Jurisdictions in the State supplying fire and rescue resources that are called into service in these incidents and that meet the time-in-service and availability criteria spelled out in the agreement are eligible for reimbursement at a rate (as of January 1990) of 519 dollars per person per 24-hour shift.

Resource Mobilization

Mobilization of mutual aid resources under the California system is a phased process. Requested assistance – personnel, equipment, or apparatus – is enlisted first from neighboring communities within the operational area, then from within the region and then inter-regionally. Each escalation in the process is designed to build on what is already there, rather than to introduce a whole new hierarchy.

Each department makes its own decision about if and how many resources it can afford to commit to mutual aid on any given incident. The decision must take into account what measures will be required to compensate for the departure of the resources they commit. In some cases, this only may involve a minimal adjustment in regular operations. In others, departments may have to activate local mutual aid agreements with neighboring communities in their own counties, call in personnel for extra shifts, or even hire extra personnel to fill in.

Resources can be dispatched singly, in strike teams, or in task forces. **Single resources** are exactly what the term implies, an individual crew, one helicopter, a single engine, etc.

A **strike team** consists of a set number of the same kind of resources with a leader. An engine strike team, for example, consists of five engines, each with a three- to four-person crew, and a chief officer who serves as the strike team leader.

A **task force** is any combination of resources temporarily assembled to perform a specific mission. A task force might be three engines and two ladder trucks and their respective crews with a chief officer who serves as the task force leader.

As single resources, strike teams, or task forces are mobilized, they are assigned a radio frequency to use. In addition, each is assigned a three-letter designator and a number that identifies where the resources came from and what type of resources they are. Their respective numbers become part of the radio call "handle" assigned to them. This enables incident command staff to keep track of and assign resources appropriately.

FIRESCOPE

The 1970 fire season, in which more than 500,000 acres in Southern California burned, more than 700 homes were destroyed and 16 lives were lost, stimulated a review and overhaul of the way in which resources are coordinated and used in major wildland incidents. This process, known as FIRESCOPE (Firefighting Resources of Southern California Organized for Potential Emergencies), also contributed to major improvements to the California Fire and Rescue Mutual Aid System overall.

FIRESCOPE was chartered in 1972. The initial focus was on the wildland fire problem, and Congress directed the USFS to help Southern California fire agencies with the review.

The principal partners in FIRESCOPE were the USFS, CDF, Los Angeles County Fire Department, Los Angeles City Fire Department, Ventura County Fire Department, Santa Barbara County Fire Department, and the California OES.

Federal funding support for FIRESCOPE began in the 1972 appropriations cycle. Funding continued through the USFS throughout the planning and development of the project. FIRESCOPE was began to be implemented in 1977, and Federal funding support, much of which went for the USFA to train for departments in new systems developed under the project, continued through 1982.

The State and partner agencies then assumed most of the costs of operation and maintenance of the project. Orange County Fire Department became a partner agency in the project in 1984, and all participants continue to be active today.

FIRESCOPE's initial missions were to improve fireground operations, increase the effectiveness of the fire protection agencies, and improve multi-agency coordination. Specific goals were:

- a standardized terminology

- a system for multi-agency communications

- a coordinated process for allocating Regional resources for use in major incidents

- improved methods for forecasting fire behavior

- training of personnel from all participating agencies

The FIRESCOPE project concepts were expanded in 1985 to Northern California in a project called the California Fire Information and Resource Management System. The two projects joined in 1988 under the FIRESCOPE banner and, instead of focusing on wildland fires, took on more of an all-hazard orientation.

FIRESCOPE consists of two major programs – the multi-agency coordination system (MACS) and the Incident Command System (ICS). Both have become integral parts of the California Fire and Rescue Mutual Aid System. Both are based on an inter-agency, multi-disciplinary approach to managing problems.

MULTI-AGENCY COORDINATION SYSTEM (MACS)

MACS is the managerial aspect of FIRESCOPE. In non-emergency times, the system functions via a multi-level "decision process," a method for involving fire service personnel at all levels through-out the State in continuous inter-agency coordination. This process is used to develop universal emergency plans and procedures and address operational and technical issues that cut across agencies/jurisdictions.

Some specialized tools have been developed to support MACS day-to-day and during emergencies. For example, a common mapping system has been developed to reduce confusion caused when different responding jurisdictions are working from different map references. Fixed and mobile facilities for using infrared monitoring information from specially equipped aircraft are available to establish the incident location and perimeter, and computerized fire modeling, as well as weather and other databases, are accessible.

California OES serves as the Executive Coordinator for MACS. On a day-to-day basis, these functions are carried out at OES' Operations Coordination Center in Riverside, California, but during emergencies MACS functions can be carried out in various facilities throughout the State depending on coordination needs.

During emergencies, the inter-agency relationships, plans, and procedures developed day-to-day become the foundation for coordinated, efficient, effective allocation of available resources through the Fire and Rescue Mutual Aid System. This foundation makes it possible for representatives of each agency involved in an incident (who are assigned the responsibility and given the authority to commit their departments'/agencies' resources) to retrieve the information they need about the incident, determine resource needs for effective response, prioritize those needs, and allocate available resources accordingly. They can resolve issues by talking the problems out "face-to-face" – sitting around a table or on a conference call. Their familiarity with the system and with each other virtually eliminates jurisdictional biases from unfairly influencing the priority attached to allocation of certain resources.

INCIDENT COMMAND SYSTEM (ICS)

The ICS is an organizational management structure that allows diverse organizations/agencies to respond and work together using standard terminology, uniform procedures, and compatible communications. The ICS concept is based on universal agreement to use a unified command structure. This means that all agencies in a multi-jurisdictional emergency play a part in determining overall incident response objectives; selecting strategies for achieving those objectives; planning and conducting integrated tactical operations; and making maximum use of all assigned resources.

California OES maintains a radio system that provides for effective communication support for resource mobilization and emergency mutual aid operations under the ICS. The system can be utilized for local, area, inter-area, Regional, Inter-Regional, and Statewide communications. The system operates on four cross band radio frequencies to provide two operating channels. Twenty-two mountain top mobile repeaters, each serving a single channel, are used to relay signals. Using the repeaters in combination with the State's microwave system makes it possible to link Regional and operational area dispatch centers throughout the State.

Framers of the ICS identified a full range of functional roles – e.g., Incident Command, information dissemination, planning, operations, logistics, finance, safety monitoring, etc. – to be carried out in emergencies and constructed an organizational structure which encompasses all those roles. (See Appendix D.) In practice, the organizational structure for command at any incident depends on the type and size of the incident itself, with responsibility for all functions – strategic as well as tactical operations – vested in the Incident Commander. In some incidents, the Incident Commander and one or two staff may be adequate to perform all necessary functions. Other situations, particularly those involving large areas and several agencies/jurisdictions, may require a much larger organizational structure, with a different person or group assigned to carry out each function.

The ICS has been used in California in a wide range of emergencies – not only those caused by fires but also those resulting from floods, earthquakes, riots, and other natural and man-made incidents. In the ICS structure, the functional responsibility for resource needs determination and allocation rests with the operations chief. The operations chief determines what resources are needed for effective response based on the overall objectives and the Incident Action Plan. Those requirements are then passed on to the logistics section chief who supervises the ordering of resources. The logistics section looks first to the local resource inventory to meet requirements identified by the operations chief.

When local resource inventories are exhausted, or in the event that unavailable specialized personnel or equipment is needed, a mutual aid request is sent to the next level in the Mutual Aid System, the operational area coordinator. As a result, the operations chief and the logistics chief serve as the link between the ICS and the Mutual Aid System. (See Appendix D.)

Because of its functional orientation, the ICS sometimes is at variance with the traditional fire department chain-of-command concept. During an incident, a department's fire chief may not be the Incident Commander. His or her deputy or operations chief may have been trained to assume that position instead, and the fire chief may be more qualified to assume another position in the incident command organization.

An effective ICS, therefore, requires that all players be trained to function within the system as it is designed and that each person accepts its tenants. As a result, the system is most effective when it is used day-to-day as well as in major incidents – from small, single department situations to big multi-agency emergency operations.

THE DEPARTMENT OF FORESTRY AND FIRE PROTECTION'S ROLE

One factor that has expedited the implementation of enhancements, particularly ICS, to California's Fire and Rescue Mutual Aid System is the unique fire protection role of the State's CDF. A major portion of the State, 32 million acres, is State-responsibility land for which CDF provides fire protection. (See Appendix E.)

In addition, CDF provides first-response fire protection under contract to 33 counties, which include some 26 incorporated cities, in the State. That is, CDF is responsible for structural firefighting, as well as wildland fire protection. In some cases, this means that CDF personnel staff the entire department. In others, CDF staffs only some stations, and in still others, only some positions in some stations.

CDF also provides fire protection to nearly 4 million acres of Federal land within the State under agreements with the USFS, the Bureau of Land Management, and other Federal agencies and back-up assistance on request for incidents on any of the more than 44 million additional acres of Federal land within the State.

In effect, CDF is like a State-wide fire department. Any new systems and procedures it adopts automatically have a broad implementation reach, and since wildland fires are a fairly regular occurrence in the State, CDF has ample opportunity to demonstrate the utility of these systems and procedures. As a result, its adoption of the ICS and its participation in the Fire and Rescue Mutual Aid System, have been major factors in leading other departments to do the same.

HOW THE MUTUAL AID SYSTEM IS USED

THE "PAINT" FIRE

Over the years, the California Fire and Rescue Mutual Aid System has been used most often on major wildland fire incidents. As drought conditions continue in the State, the 1991 fire season is likely to provide more opportunities for the system to be exercised.

A recap of the "Paint" fire in June 1990, the most destructive wildland fire in California's history, illustrates not only how the State's Mutual Aid System works, but also its importance to effective management of major incidents.

Santa Barbara County and much of Southern California had suffered drought conditions for several years. That, combined with the Santa Barbara's penchant for very hot and dry "Santa Ana" and "Sundowner" winds, had aggravated fire-prone conditions in the area. As a consequence, a "Red Flag Alert" was in effect. A "Red Flag Alert" means that, based on weather conditions and other factors, the danger of fire is real.

The Santa Barbara County Fire Department had already begun to gear up for possible mobilization, and its coordination center had been activated in response to the "Red Flag Alert." During "Red Flag Alert" periods, the coordination center serves as the focal point for inter-agency coordination and processing of mutual aid requests in Region I (the mutual aid region in which Santa Barbara County is located).

The "Paint" fire was first reported, via a 9-1-1 call, at 6:02 p.m., on June 27. A Santa Barbara County brush engine responded immediately, arriving at 6:05 p.m. The fire had begun on Painted Cave Road, just south of the intersection with Highway 154, in the Los Padres National Forest. It had been set by an arsonist in a steep drainage area. The fire had moved down the canyon quickly, fanned by the wind and fueled by the heavy chaparral fuel and dry weather conditions. The responding crew found a fire two to five acres in size.

Major spotting was reported at 6:10 p.m. This means that wind-carried embers and other burning material were starting new fires away from the main blaze.

The Los Prietas Hot Spots, a specially trained wildland fire team of the USFS, arrived on scene at the same time, and additional forces were dispatched by both Santa Barbara County and the USFS.

The Santa Barbara County Fire Department and USFS quickly assessed the fire situation and established joint command. By 6:20 p.m., an incident commander was assigned from the fire department, and the Forest Service became the focal point for obtaining mutual aid resources required.

The ICS, described in more depth later in this report, is a major element in California's Mutual Aid System. Under the ICS, the operations section chief and the logistics section chief serve as the focal points for determining resource needs and ordering those resources through the system.

At 6:27 p.m., the Santa Barbara County Fire chief reported 50 to 60 mph winds and major spotting heading toward homes and other structures in the area. Limiting the lateral spread of the fire north and south, protecting the structures in its path and evacuating as necessary became the major objectives of the response. The logistics section asked for help through the Fire and Rescue Mutual Aid System in the form of engine strike teams, bulldozers, and water tenders. They came, primarily from fire departments in neighboring counties in the region, as well as from the CDF and USFS.

Nineteen strike teams of engines responded to Santa Barbara and can be credited with saving many structures. They responded rapidly, and their home departments took the steps necessary to compensate for their departure. In some cases, this only may have involved a minimal adjustment in regular operations. In others, the home departments may have had to call on local mutual aid agreements with neighboring departments in their own counties, or even hire extra personnel to fill in for the resources contributed to fight the "Paint" fire.

This response is particularly notable because there were other major wildland fires burning in California at the time of the "Paint" fire. The State Fire and Rescue Mutual Aid System was faced with the added challenge of mustering, distributing and, in some cases, re-distributing resources among several locations over an extended period of time.

For example, a major wildland incident was in progress in Orange County, about 100 miles southeast of Santa Barbara. The Orange County Fire Department was already drawing on mutual aid resources from other departments in Southern California. When a request for mutual aid assistance for the "Paint" incident was received, however, Orange County made a decision to help. It contributed some of its own strike teams, through the State Fire and Rescue Mutual Aid System, to the Santa Barbara response. Before they could be sent home, some of these same strike teams were re-assigned again through the Mutual Aid System when wildland fires broke out some days later in Northern California near Redding.

By 6:45 p.m., on June 28, the fire had crossed the boundary into the Los Padres National Forest. Firefighting operations, including air tanker drops over the fire and structure protection activities continued through the night and the following day.

By 8:00 a.m., on June 29, the fire was 4,200 acres in size, but the rate of spread and intensity had lessened. Weather was milder, and humidity was rising. An estimated 520 residential structures and at least two shopping centers had been lost. By late afternoon on June 30 the fire had been contained.

The final toll of this record-breaking incident was one person dead, more than 600 structures and 4,900 acres burned, and 400 million dollars in damages. The Fire and Rescue Mutual Aid System had contributed more than 100 engines and bulldozers, as well as the necessary personnel to operate them, multiplying tenfold the size of the firefighting force available in this incident.

NON-FIRE INCIDENTS

The California Fire and Rescue Mutual Aid System was designed to be used in emergencies caused by a variety of hazards. Over the years, the system has been called on to provide resources for emergency response in floods, the Whittier and Coalinga earthquakes, and riots and non-emergency assistance during the Olympic Games in Los Angeles and the visit of the Pope during the late 1970s.

The Loma Prieta Earthquake, which struck the San Francisco and Monterey Bay area in October 1989, provided another major example of the California system in action. (See Appendix G.)

The earthquake, measuring 7.1 on the Richter scale and centered just north of Santa Cruz, California, occurred October 17 and lasted only about 15 seconds. In that brief period, however, it damaged or destroyed more than 100,000 buildings and left 7,000 people homeless. It claimed 63 lives, caused more than 2,400 injuries, and resulted in many billions of dollars in damages.

Fire and Rescue Mutual Aid System coordinators at the local and Regional levels, as well as the State coordination facilities operated by the California OES, received hundreds of other requests for mutual aid assistance following the quake. As the coordinators at each level exhausted their resource inventories – personnel, equipment, and facilities – they sent requests up to the next level to be filled.

In all, 17 engine strike teams, three OES heavy rescue units (one of which was flown in from Orange County in Southern California), three OES mobile supply units, three local rescue units a lighting unit and 15 Alameda County engines and truck companies were dispatched through the Fire and Rescue Mutual Aid System. These resources, as well as other types of assistance, were dispatched to help with a variety of emergency response tasks. For example, personnel from the California State Fire Marshal's Office joined teams charged with inspecting health care facilities, schools, and other public buildings in the affected area for fire and life safety hazards.

On October 19, the Oakland Fire Department requested a major incident management team through the Fire and Rescue Mutual Aid System for assistance on the Cypress incident, the collapse of a one-and-one-quarter-mile stretch of the upper deck of the Interstate 880 freeway onto the lower level. The incident had buried 58 vehicles, claimed 42 lives, and injured 348 others. The Oakland Fire Department had been working the incident for about 36 hours and personnel were exhausted.

Their request was relayed through the system and within a few hours an inter-agency management team, composed of California CDF and USFS personnel, was assembled to fill the key command and support roles in the Incident Command Structure. The team stayed with the incident through October 23 when the last rescue was performed and demobilization operations were begun.

Numerous fire departments from unaffected areas called in off-duty personnel who reported to fire stations and staging areas to be available for dispatch. In addition, CDF committed 21 strike teams of engines, three strike teams of bulldozers, nine prison inmate fire crews, six prison inmate kitchen crews, and more than 100 other CDF personnel to respond to incidents in areas for which CDF has fire protection responsibility.

Two engines from CDF's Hollister, California, station provided assistance to the city of Hollister, followed by two CDF engine strike teams, one each from CDF's San Benito/Monterey and San Luis Obispo Ranger Units. CDF resources responded to approximately 57 earthquake related fire and non-fire incidents in the city of Hollister and 22 incidents in other parts of the San Benito/Monterey areas.

WHY CALIFORNIA'S MUTUAL AID SYSTEM WORKS

A number of factors have contributed to the success of the California Fire and Rescue Mutual Aid System over the years. They may provide helpful lessons for other States or regions interested in improving their own mutual aid systems.

- Basic attitudinal changes have been required to make the California system function successfully. Many in the fire service have modified their "we take care of our own" notions in the face of the economic and social realities of the day. The willingness to give and accept help from one's neighbor is essential to survival and the underpinning of the mutual aid system in the State.

- Attitudes about the chain-of-command concept have had to change as well. Chain of command has always been very important in para-military organizations like fire and police agencies. However, much of the fire service in California seems to have accepted that effective, efficient emergency response depends in large part on being able to apply the most qualified, most experienced, best trained person to each individual problem. This requires disregarding in some cases, the chain of command. In some emergency response organizations in California there may be several individuals trained as Incident Commanders. During an incident a department's fire chief may not be the Incident Commander. His or her deputy or operations chief may have been trained to assume that position instead, and the fire chief may be more qualified to assume another position in the Incident Command organization.

- Successful coordination and cooperation requires trust. And trust is built through working together over time. The multi-agency coordination system and the decision process developed as part of FIRESCOPE provide ample opportunity for a variety of jurisdictions and agencies to work together consistently to solve problems. Consequently, during an incident their representatives can approach resource allocation issues more objectively. Their familiarity with the system and with each other virtually eliminates jurisdictional biases from unfairly influencing the priority attached to allocation of certain resources, even when a number of incidents are being considered.

- Effective, efficient emergency and disaster response require the participation of a variety of disciplines – fire, police, emergency medical services, public works, etc. The California fire service's participation and acceptance of the Fire and Rescue Mutual Aid System has served as an example, bringing other disciplines together and focusing their attention on emergency response planning. This can only serve to benefit overall emergency preparedness and response capabilities in the State.

- California's mutual aid system structure has been around a long time, long enough for most firefighters and officers to have seen it work and developed a level of comfort with it. Everybody participates and even many non-fire agencies in the State are familiar enough with how the system works to be able to interface with it when necessary.

- The California system gets used repeatedly throughout the year to combat fires, particularly wildland fires. This consistent exercising of the system keeps it honed and ready for use in other emergencies.

- There is a commitment in the State to training to ensure that all potential players know how to function within the system. The concepts and procedures which support the California Fire and Rescue Mutual Aid System are an integral part of firefighter training in the State, not only in individual departments but also in community colleges and throughout the State fire training system coordinated by the State Fire Marshal's Office.

- The ICS, a concept basic to today's Fire and Rescue Mutual Aid System in California, has become a way of life for much of the fire service in the State and many non-fire agencies as well. Large, multi-jurisdictional incidents generally do not require "switching" to another mode of operation. This helps maintain smooth operations and makes each escalation easier.

- The FIRESCOPE Project is a testament to the State's commitment to improvement in its mutual aid capabilities. Since the cessation of Federal funding in 1982, the fire service in California has continued to work voluntarily to implement the project Statewide. Committees and task forces responsible for specific aspects of the project meet regularly, at their own expense, to review progress, update assignments, and consider new issues.

- Although it is exercised most often on fire incidents, the system has shown itself to be adaptable to floods and mudslides, the Olympics, a visit from the Pope, the Loma Prieta Earthquake, and other non-fire emergencies.

- The CDF, the USFS, and the U.S. Bureau of Land Management share fire protection responsibilities for well over 69 million acres of public and private land in California. In addition, CDF provides first-alarm protection, structural as well as wildland, under contract to a number of individual California communities. The pervasive presence of CDF and the Forest Service, both of which use the ICS, has been a major factor in its acceptance throughout the State. In addition, because CDF is the community fire department in many areas, they have been able to train local firefighters, right along with their own people, to use the system. Since ICS is now an integral element of the California Fire and Rescue Mutual Aid System, this has benefited the system overall.

Appendices

A. California Disaster and Civil Defense Master Mutual Aid Agreement

B. Map of California Mutual Aid Regions

C. The "Five-Party Agreement"
   Cooperative Agreement between California OES; CDF; USFS, Pacific
   Southwest Region; U.S. Bureau of Land Management, California
   Office; and U.S. National Park Service, Western Region

D. ICS and Link Between ICS and Mutual Aid System

E. Map of Wildland Fire Protection Responsibility Areas

F. "Paint" Fire Area Map

G. Loma Prieta Earthquake: Structural Collapse of I-880 and Status Map
   of Trapped Vehicles

H. Bibliography

Appendix A

California Disaster and Civil Defense Master Mutual Aid Agreement

CALIFORNIA DISASTER AND CIVIL DEFENSE
MASTER MUTUAL AID AGREEMENT

This agreement made and entered into by and between the STATE OF CALIFORNIA, its various departments and agencies, and the various political subdivisions, municipal corporations, and other public agencies of the State of California;

W I T N E S S E T H :

WHEREAS, it is necessary that all of the resources and facilities of the State, its various departments and agencies, and all its political subdivisions, municipal corporations, and other public agencies be made available to prevent and combat the effect of disasters which may result from such calamities as flood, fire, earthquake, pestilence, war, sabotage, and riot: and

WHEREAS, it is desirable that each of the parties hereto should voluntarily aid and assist each other in the event that a disaster should occur, by the interchange of services and facilities, including, but not limited to, fire, police, medical and health, communication, and transportation services and facilities, to cope with the problems of rescue, relief, evacuation, rehabilitation, and reconstruction which would arise in the event of a disaster; and

WHEREAS, it is necessary and desirable that a cooperative agreement be executed for the interchange of such mutual aid on a local, countywide, regional, statewide, and interstate basis;

1.   Each party shall develop a plan providing for the effective mobilization of all its resources and facilities, both public and private, to cope with any type of disaster.

2.   Each party agrees to furnish resources and facilities and to render services to each and every other party to this agreement to prevent and combat any type of disaster in accordance with duly adopted mutual aid operational plans, whether heretofore or hereafter adopted, detailing the method and manner by which such resources, facilities, and services are to be made available and furnished, which operational plans may include provisions for training and testing to make such mutual aid effective; provided, however, that no party shall be required to deplete unreasonably its own resources, facilities, and services in furnishing such mutual aid.

3.   It is expressly understood that this agreement and the operational plans adopted pursuant thereto shall not supplant existing agreements between some of the parties hereto providing for the exchange or furnishing of certain types of facilities and services on a reimbursable, exchange, or other basis, but that the mutual aid extended under this agreement and the operational plans adopted pursuant thereto, shall be without reimbursement unless otherwise expressly provided for by the parties to this agreement or is provided in Sections 1541, 1586, and 1587, Military and Veterans Code; and that such mutual aid is intended to be available in the event of a disaster of such magnitude that it is, or is likely to be, beyond the control of a single party and requires the combined forces of several or all of the parties to this agreement to combat.

4.   It is expressly understood that the mutual aid extended under this agreement and the operational plans adopted pursuant thereto shall be available and furnished in all cases of local peril or emergency and in all cases in which a *STATE OF EXTREME EMERGENCY* has been proclaimed.

Revised 09/88

5.   It is expressly understood that any mutual aid extended under this agreement and the operational plans adopted pursuant thereto, is furnished in accordance with the "California Disaster Act" and other applicable provisions of law, and except as otherwise provided by law that: "The responsible local official in whose jurisdiction an incident requiring mutual aid has occurred shall remain in charge at such incident including the direction of such personnel and equipment provided him through the operation of such mutual aid plans." (Sec. 1564, Military and Veterans Code.)

6.   It is expressly understood that when and as the State of California enters, into mutual aid agreements with other states and the Federal Government, the parties to this agreement shall abide by such mutual aid agreements in accordance with the law.

7.   Upon approval or execution of this agreement by the parties hereto all mutual aid operational plans heretofore approved by the State Disaster Council, or its predecessors, and in effect as to some of the parties hereto, shall remain in full force and effect as to them until the same may be amended, revised, or modified. Additional mutual aid operational plans and amendments, revisions, or modifications of existing or hereafter adopted mutual aid operational plans, shall be adopted as follows:

a.   Countywide and local mutual aid operational plans shall be developed by the parties thereto and are operative as between the parties thereto in accordance with the provisions of such operational plans. Such operational plans shall be submitted to the State Disaster Council for approval. The State Disaster Council shall notify each party to such operational plans of its approval, and shall also send copies of such operational plans and who are in the same area and affected by such operational plans. Such operational plans shall be operative as to such other parties 20 days after receipt thereof unless within that time the party by resolution or notice given to the State Disaster Council, in the same manner as notice of termination of participation in this agreement, declines to participate in the particular operational plan.

b.   Statewide and regional mutual aid operational plans shall be approved by the State Disaster Council and copies thereof shall forthwith be sent to each and every party affected by such operational plans. Such operational plans shall be operative as to the parties affected thereby 20 days after receipt thereof unless within that time the party by resolution or notice given to the State Disaster Council, in the same manner as notice of termination of participation in this agreement, declines to participate in the particular operational plan.

c.   The declination of one or more of the parties to participate in a particular operational plan or any amendment, revision or modification thereof, shall not affect the operation of this agreement and the other operational plans adopted pursuant thereto.

d.   Any party may at any time by resolution or notice given to the State Disaster Council, in the same manner as notice of termination of participation in this agreement, decline to participate in any particular operational plan, which declination shall become effective 20 days after filing with the State Disaster Council.

e.   The State Disaster Council shall send copies of all operational plans to those state departments and agencies designated by the Governor. The Governor may, upon behalf of any department or agency, give notice that such department or agency declines to participate in a particular operational plan.

f.   The State Disaster Council, in sending copies of operational plans and other notices and information to the parties to this agreement, shall send copies to the Governor and any department or agency head designated by him; the chairman of the board of supervisors, the clerk of the board of supervisors, the County Disaster Council, and any other officer designated by a county; the mayor, the clerk of the city council, the City Disaster Council, and any other officer designated by a city; the executive head, the clerk of the governing body, or other officer of other political subdivisions and public agencies as designated by such parties.

8.   This agreement shall become effective as to each party when approved or executed by the party, and shall remain operative and effective as between each and every party that has heretofore or hereafter approved or executed this agreement, until participation in this agreement is terminated by the party. The termination by one or more of the parties of its participation in this agreement shall not affect the operation of this agreement as between the other parties thereto. Upon approval or execution of this agreement the State Disaster Council shall send copies of all approved and existing mutual aid operational plans affecting such party which shall become operative as to such party 20 days after receipt thereof unless within that time the party by resolution or notice given to the State Disaster Council, in the same manner as notice of termination of participation in this agreement, declines to participate in any particular operational plan. The State Disaster Council shall keep every party currently advised of who the other parties to this agreement are and whether any of them has declined to participate in any particular operational plan.

9.   Approval or execution of this agreement shall be as follows:

a.   The Governor shall execute a copy of this agreement on behalf of the State of California and the various departments and agencies thereof. Upon execution by the Governor a signed copy shall forthwith be filed with the State Disaster Council.

b.   Counties, cities, and other political subdivisions and public agencies having a legislative or governing body shall by resolution approve and agree to abide by this agreement, which may be designated as *"CALIFORNIA DISASTER AND CIVIL DEFENSE MASTER MUTUAL AID AGREEMENT."* Upon adoption of such a resolution, a certified copy thereof shall forthwith be filed with the State Disaster Council.

c.   The executive head of those political subdivisions and public agencies having no legislative or governing body shall execute a copy of this agreement and forthwith file a signed copy with the State Disaster Council.

10.   Termination of participation in this agreement may be effected by any party as follows:

a.   The Governor on behalf of the State and its various departments and agencies, and the executive head of those political subdivisions and public agencies having no legislative or governing body, shall file a written notice of termination of participation in this agreement with the State Disaster Council and this agreement is terminated as to such party 20 days after the filing of such notice.

b.   Counties, cities, and other political subdivisions and public agencies having a legislative or governing body shall by resolution give notice of termination of participation in this agreement and file a certified copy of such resolution with the State Disaster Council, and this agreement is terminated as to such party 20 days after the filing of such resolution.

IN WITNESS WHEREOF this agreement has been executed and approved and is effective and operative as to each of the parties as herein provided.

/s/ EARL WARREN
GOVERNOR

On behalf of the State of California and all its Departments and Agencies.

(SEAL)   ATTEST:

November 15, 1950

/s/ FRANK M. JORDAN
SECRETARY OF STATE

Revised 09/88

Appendix B

Map of California Mutual Aid Regions

State of California
OFFICE OF EMERGENCY SERVICES

MUTUAL AID REGIONS

Revised 09/88

Appendix C

The "Five-Party Agreement"

Cooperative Agreement between California OES; CDF; USFS, Pacific Southwest Region; US. Bureau of Land Management, California Office; and U.S. National Park Service, Western Region

COOPERATIVE AGREEMENT
between
STATE OF CALIFORNIA, OFFICE OF EMERGENCY SERVICES;
STATE OF CALIFORNIA, DEPARTMENT OF FORESTRY AND FIRE PROTECTION
PACIFIC SOUTHWEST REGION, USDA FOREST SERVICE;
USDI BUREAU OF LAND MANAGEMENT, CALIFORNIA OFFICE; and
USDI NATIONAL PARK SERVICE, WESTERN REGION

THIS AGREEMENT made and entered into on July 1, 1990, by and between the State of California, Office of Emergency Services, hereinafter referred to as State OES; the USDA Forest Service, Pacific Southwest Region; the State of California, Department of Forestry and Fire Protection; the USDI Bureau of Land Management, California Office; and the USDI National Park Service, Western Region, the latter four parties hereinafter referred to as Forest Agencies, under the provisions of the Act of December 12, 1975, PL 94-148, the Act of April 24, 1950 (16 USC 572), 42 USC 1856, and the Disaster Relief Act of 1974, PL 93-288.

RECITALS

1. The Forest Agencies are responsible for providing a level of wildland fire protection for both federal and non-federal lands within the state of California; and

2. The State OES is responsible for the management and coordination of the State Fire and Rescue Mutual Aid System; and

3. The Forest Agencies, at times of severe wildfire conditions, often have need of emergency apparatus to provide structural fire protection and to supplement their respective agency-controlled resources to combat the effects of fire; and

4. The State OES and/or various local government jurisdictions have such emergency apparatus which can be made available, in the spirit of cooperation, for dispatch and use through the State Fire and Rescue Mutual Aid System; and

5.
   It is desirable that the State OES and the Forest Agencies establish and enter into an agreement for the prudent use of, and reimbursement for, services of such emergency apparatus and personnel;

THEREFORE, it is agreed as follows:

DEFINITIONS

6. **LOCAL JURISDICTION** shall mean any subdivision of government, including agencies or institutions to which the State OES has, through agreement, assigned State OES-owned emergency apparatus; or who provide locally-owned resources under provisions of the State Fire and Rescue Mutual Aid System.

7. **EMERGENCY APPARATUS** (hereinafter called apparatus) shall mean any vehicular apparatus provided through the State Fire and Rescue Mutual Aid System.

8. **EMERGENCY PERSONNEL** shall mean any personnel responding on or with emergency apparatus.

9. **MOBILIZATION CENTER** shall mean an off-incident location at which emergency apparatus and personnel are temporarily located pending assignment, release or reassignment.

10. **DEMOBILIZATION CENTER/FACILITY** shall mean that location or facility established at or near an incident for the processing of apparatus and personnel prior to release to its home base.

11. **STAGING AREA** shall mean the location where apparatus and personnel are assigned to an incident for deployment on a three-minute availability status.

TERMS AND CONDITIONS

REQUESTS FOR AND RELEASE OF EMERGENCY APPARATUS AND PERSONNEL

13. Forest Agency requests for apparatus and personnel shall be placed to the State Fire and Rescue Mutual Aid System following procedures set forth in the California Fire Service and Rescue Emergency Mutual Aid Plan.

    Forest Agencies and State OES will use the Resource Order Form (Form MACS 420 and 420A or equivalent) for all requests. Forest Agencies shall not be responsible for any apparatus and personnel not confirmed by their respective order and request number(s).

14. Forest Agency release or reassignment of apparatus used on wildland emergencies will be coordinated through the on-scene OES Fire and Rescue officer, the local jurisdiction agency representative, or their authorized representative. Such OES officer or representative will be responsible for the inspection and inventory of such apparatus prior to release to its home base.

PROTECTIVE CLOTHING AND EQUIPMENT

15. It shall be the responsibility of the jurisdiction sending emergency personnel to insure that such personnel are provided protective clothing and equipment as required by California Code of Regulations, Title 8, Section 3410, Article 10.1, Section 3401, et seq.

REIMBURSEMENT

16. Provisions and procedures for reimbursement by Forest Agencies for fire suppression assistance are defined separately for State OES-owned apparatus and personnel and locally-owned apparatus and personnel on Exhibits 1 and 2, respectively, attached hereto and made a part hereof. (Reimbursement for personnel on State OES-owned apparatus, in most cases shall be to "local jurisdictions" who provide such personnel by separate agreement with State OES.)

OFFICIALS NOT TO BENEFIT

17. No member of, or Delegate to Congress or Resident Commission shall be admitted to any share or part of this agreement or to any benefit to arise therefrom, unless it is made with a corporation for its general benefit.

PREVIOUS AGREEMENTS CANCELED

18. This agreement supersedes and cancels the Memorandum of Understanding entered into on July 1, 1987 between State OES; Pacific Southwest Region, USDA Forest Service; and State of California Department of Forestry and Fire Protection.

Cooperative Agreement Between OES/CDF/USFS/BLM/NPS

CIVIL RIGHTS

19.   The extension of benefits under the provisions of this agreement shall be without discrimination as to age, handicap, race, color, creed, sex, or national origin.

TERMINATION

20.   This agreement shall remain in effect until June 30, 1993. It may be terminated by any one of the parties hereto upon thirty (30) days notice, in writing, to the other parties.

APPROPRIATED FUND LIMITATION

21.   Nothing herein shall be interpreted as obligating any parties herein to expend funds or as involving the United States or the State of California in any contract or other obligation for the future payment of money in excess of appropriations authorized by law and administratively allocated for the work contemplated in this agreement.

IN WITNESS WHEREOF, the parties hereto have executed this agreement

STATE OF CALIFORNIA
OFFICE OF EMERGENCY SERVICES

By: _____   _June 20, 1990_
          Director   Date

STATE OF CALIFORNIA
DEPARTMENT OF FORESTRY AND FIRE PROTECTION

By: _____   _7/2/90_
          Director   Date

USDA FOREST SERVICE
PACIFIC SOUTHWEST REGION

By: _____   _July 13, 1990_
          Regional Forester   Date

USDI BUREAU OF LAND MANAGEMENT
CALIFORNIA STATE OFFICE

By: _____   _7/5/90_
          California State Director   Date

USDI NATIONAL PARK SERVICE
WESTERN REGION

By: _____   _7/18/91_
          Regional Director   Date

By: _____   JUL 1 9 1990
          Regional Contracting Officer   Date

Cooperative Agreement Between OES/CDF/USFS/BLM/NPS

EXHIBIT 1

REIMBURSEMENT PROCEDURES - PERSONNEL ON STATE OES-OWNED APPARATUS

The following procedures are for the use of personnel on State OES-owned emergency and associated support apparatus. Terms established in this section shall be made binding upon local jurisdictions by State OES and shall not be subject to interpretation or rejection by the jurisdiction providing assistance.

It is understood and agreed that a local jurisdiction providing personnel on State OES-owned apparatus shall, if it so desires, obtain reimbursement for such response by billing the using Forest Agency in accordance with the following:

1. There shall be no reimbursement for response of personnel on State OES-owned apparatus of less than 12 hours duration. However, should State OES-owned apparatus be requested for assignment to a mobilization center for standby duty, the reimbursement period shall begin with the time of initial dispatch of said apparatus from its assigned home base. Additionally, there shall be only one 12-hour free period for each resource from time of original dispatch, regardless of numbers of assignments or Forest Agencies committing that resource, until its return to home base.

2. The reimbursement formula shall be based upon an average combined overtime salary of one Captain, one Engineer, and one Firefighter, using the six highest paid departments, as determined by State OES, in each of the six mutual aid regions in California (total 36). The formula shall utilize the average daily overtime salary of the 36 representative positions (i.e., the sum of the 36 daily overtime salaries divided by 36). Such formula to be reviewed, evaluated and adjusted by OES on the first day of January of each year. All data calculations shall be subject to audit by the Forest Agencies.

   This rate of payment shall constitute full reimbursement, including direct and indirect costs, to jurisdictions relative to personnel provided. Current rate, effective <u>January 1, 1990</u>, is established at <u>$519</u> per person per 24-hour shift.

   Reimbursement is for work which requires 24-hour availability without regard to calendar days. Reimbursement for fractional days shall be 50 percent of the current 24-hour rate for each 12 hours or fraction thereof.

3. Reimbursement shall be made directly to the jurisdiction providing the apparatus, and not to individual crew members, support, or coordinating personnel.

4. Crew size on engine companies responding to Forest Agency requests shall not be less than three (3). Forest Agencies may specify a 4th person at the time of request. Reimbursement shall be for three persons unless the Forest Agency specifies a fourth person. Reimbursement for a fourth person shall only be provided when that person has been specifically requested by the Forest Agency. A request for ICS Type 1 engine shall imply the authorization for the fourth person. However, no assignee of OES-owned engines shall be required to ride personnel on the apparatus tailboard if it is contrary to department practice.

5. Forest Agencies shall provide reimbursement for personnel requested by the Forest Agencies to coordinate or otherwise support the State OES-owned apparatus used on incidents. Reimbursement shall be allowed for an assistant strike team leader should the sending jurisdiction choose to provide one. Reimbursement shall be at the current rate specified in item 2, above.

   Reimbursement shall be made only for such support personnel that have been specifically requested by the Forest Agency. Any support personnel not given an Order/Request number shall be considered a voluntary contribution from the sending agency and not subject to reimbursement.

   In no case will a second support vehicle be assigned to an individual strike team.

Exhibit 1   Cooperative Agreement Between OES/CDF/USFS/BLM/NPS   Rev. 6/90

6.   Forest Agencies shall reimburse local jurisdictions for use of local jurisdiction support equipment provided in conjunction with requested personnel provided in item 5, above. Reimbursement shall be calculated on a cost-per-mile basis or daily guarantee, whichever is greater, at the rate currently established by the requested Agencies for the type or category of vehicle used. Such reimbursement shall be considered as covering all costs related to use of such vehicles except as provided in item 9, below.

7.   Local jurisdictions will prepare OES Form 42 (Emergency Activity Record), which is the basis for reimbursement due and invoice preparation. These forms are provided by OES Fire and Rescue Division. The forms (F-42) must be signed by a responsible officer of the jurisdiction seeking reimbursement and forwarded to OES Fire and Rescue Division in Sacramento.

State OES will process the F-42 data into invoices (F-142) and return to the local jurisdiction for signature. Upon return, State OES will forward the invoice to the appropriate Forest agency for payment.

INVOICES NOT RETURNED TO OES BY THE LOCAL JURISDICTION WITHIN 60 DAYS OF RECEIPT BY THE LOCAL JURISDICTION WILL BE CONSIDERED NULL AND VOID.

8.   State OES shall assume operational costs, including necessary motor fuels and lubricants used in State-owned OES apparatus while responding to and returning from Forest Agency incidents. Local jurisdictions assume the same responsibility for their support equipment responding with State OES-owned apparatus. It shall be the responsibility of the sending jurisdiction to provide the necessary means of payment for such costs.

9.   Forest Agencies will provide for motor fuel and lubricants, normal servicing costs, and minor repairs incidental to operation of apparatus including local jurisdiction support equipment while under direction and control of the requesting Forest Agency. Minor repair is defined as any repair necessary to keep the equipment in operation on the fire which requires not more than two hours for one mechanic for any one job.

10.   State OES assumes the normal cost of repair of damage to State OES-owned apparatus which may result from use under the terms of this agreement. Loss or damage to local jurisdiction support equipment, including repairs due to normal wear and tear or due to negligent or inefficient operation by the operator, shall be the responsibility of the local jurisdiction providing the equipment.

Exhibit 1   Cooperative Agreement Between OES/CDF/USFS/BLM/NPS   Rev. 6/90

EXHIBIT 2
LOCAL JURISDICTION APPARATUS AND PERSONNEL

It is expressly understood and agreed that Forest Agencies and State OES cannot, and have no intent to, enter an agreement affecting reimbursement for local jurisdiction services, except as it applies to Forest Agency requests and use. Local jurisdictions who provide their mutual aid apparatus and personnel to Forest Agencies through the State Fire and Rescue Mutual Aid System, do so on a voluntary basis, and accept the following provisions for reimbursement. Assistance by hire contracts are not included in this agreement.

1. All requests for local jurisdiction apparatus and personnel by Forest Agencies shall be processed directly through the State Fire and Rescue Mutual Aid System using standard request procedures.

2. There shall be no reimbursement for response of personnel on local government apparatus of less than 12 hours duration. However, should local government apparatus be requested for assignment to a mobilization center for standby duty, the reimbursement period shall begin with the time of initial dispatch of said apparatus from its assigned home base. Additionally, there shall be only one 12-hour free period for each resource from time of original dispatch, regardless of numbers of assignments or Forest Agencies committing that resource, until its return to home base.

3. The reimbursement formula shall be based upon an average combined overtime salary of one Captain, one Engineer, and one Firefighter, using the six highest paid departments, as determined by State OES, in each of the six mutual aid regions in California (total 36). The formula shall utilize the average daily overtime salary of the 36 representative positions (i.e., the sum of the 36 daily overtime salaries divided by 36). Such formula to be reviewed, evaluated and adjusted by OES on the first day of January of each year.

   All data and calculations shall be subject to audit by the Forest Agencies.

   This rate of payment shall constitute full reimbursement, including direct and indirect costs, to jurisdictions relative to personnel provided. Current rate, effective January 1, 1990, is established at $519 per person per 24-hour shift.

   Reimbursement is for work which requires 24-hour availability without regard to calendar days. Reimbursement for fractional days shall be 50 percent of the current 24-hour rate for each 12 hours or fraction thereof.

4. Reimbursement - Fire Engines: Shall be in accordance with the current Schedule of Equipment Rates established pursuant to Section 420 of the Disaster Relief Act, PL 93-288, by the Federal Emergency Management Agency. These rates are based on engine horsepower. For example, the current rate for a Type 1 or 2 engine (based on 250 horsepower rating) is $18.00 per hour, with a 16 hour maximum allowable charge per 24 hour period. This rate or other lower rate for smaller apparatus includes all allowable charges, including mileage and pumping time.

5. Forest Agencies shall reimburse local jurisdictions for use of local jurisdiction support equipment provided in conjunction with requested personnel provided in Item 4, above. Reimbursement shall be calculated on a cost-per-mile basis or daily guarantee, whichever is greater, at the rate currently established by the requesting Agencies for the type or category of vehicle used. Such reimbursement shall be considered as covering all costs related to use of such vehicles except as provided in Paragraph Item 11, below.

Exhibit 2   Rev. 6/90

6.   Reimbursement shall be made directly to the jurisdiction providing the apparatus, and not to individual crew members, support, or coordinating personnel.

7.   Crew size on engine companies responding to Forest Agency requests shall not be less than three (3). Forest Agencies may specify a 4th person at the time of request. Reimbursement shall be for three persons unless the Forest Agency specifies a fourth person. Reimbursement for a fourth person shall only be provided when that person has been specifically requested by the Forest Agency. A request for ICS Type 1 engine shall imply the authorization for the fourth person.

8.   Forest Agencies shall provide reimbursement for personnel requested by the Forest Agencies to coordinate or otherwise support the mutual aid apparatus used on incidents. Reimbursement shall be allowed for an assistant strike team leader should the sending jurisdiction choose to provide one. Reimbursement shall be at the current rate specified in item 3, above.

      Reimbursement shall be made only for such support personnel that have been specifically requested by the Forest Agency. Any support personnel not given an Order/Request number shall be considered a voluntary contribution from the sending agency and not subject to reimbursement.

      In no case will a second support vehicle be assigned to an individual strike team.

9.   Local jurisdictions will prepare OES Form 42 (Emergency Activity Record), which is the basis for reimbursement due and invoice preparation. These forms are provided by OES Fire and Rescue Division. The forms (F-42) must be signed by a responsible officer of the jurisdiction seeking reimbursement and forwarded to OES Fire and Rescue Division in Sacramento.

      State OES will process the F-42 data into invoices (F-142) and return to the local jurisdiction for signature. Upon return, State OES will forward the invoice to the appropriate Forest Agency for payment.

10.   Local jurisdiction shall assume operational costs, including necessary motor fuels and lubricants used in its apparatus while responding to and returning form Forest Agency incidents. Local jurisdictions assume the same responsibility for their support equipment responding with State OES-owned apparatus. It shall be the responsibility of the sending jurisdiction to provide the necessary means of payment for such costs.

11.   Forest Agencies will provide for motor fuel and lubricants, normal servicing costs, and minor repairs incidental to operation of apparatus including local jurisdiction support equipment while subject to direction and control for the requesting Forest Agency. Minor repair is defined as any repair necessary to keep the equipment in operation on the fire which requires no more than two hours for one mechanic for any one job.

12.   Loss or damage repair to local jurisdiction apparatus or support equipment, including repairs due to normal wear and tear or due to negligent or inefficient operation by the operator, shall be the responsibility of the local jurisdiction providing the equipment.

13.   Locally-owned emergency apparatus cannot be transferred form one incident to another without the sending jurisdiction's approval. Approval for such redirection shall be secured by the using Forest Agency through the State Fire and Rescue Mutual Aid System.

Exhibit 2   Rev. 6/90

Appendix D

ICS

Under the ICS, the organizational structure for management of a major incident is based on the functional roles that must be carried out.

Link Between ICS and Mutual Aid System

Under the ICS, the operations section chief is charged with determining needs for equipment, apparatus, and personnel adequate for the response effort. Ordering of these resources through the Fire and Rescue Mutual Aid System is the responsibility of the logistics section chief.

INCIDENT COMMAND SYSTEM (ICS)
MAJOR INCIDENT ORGANIZATION

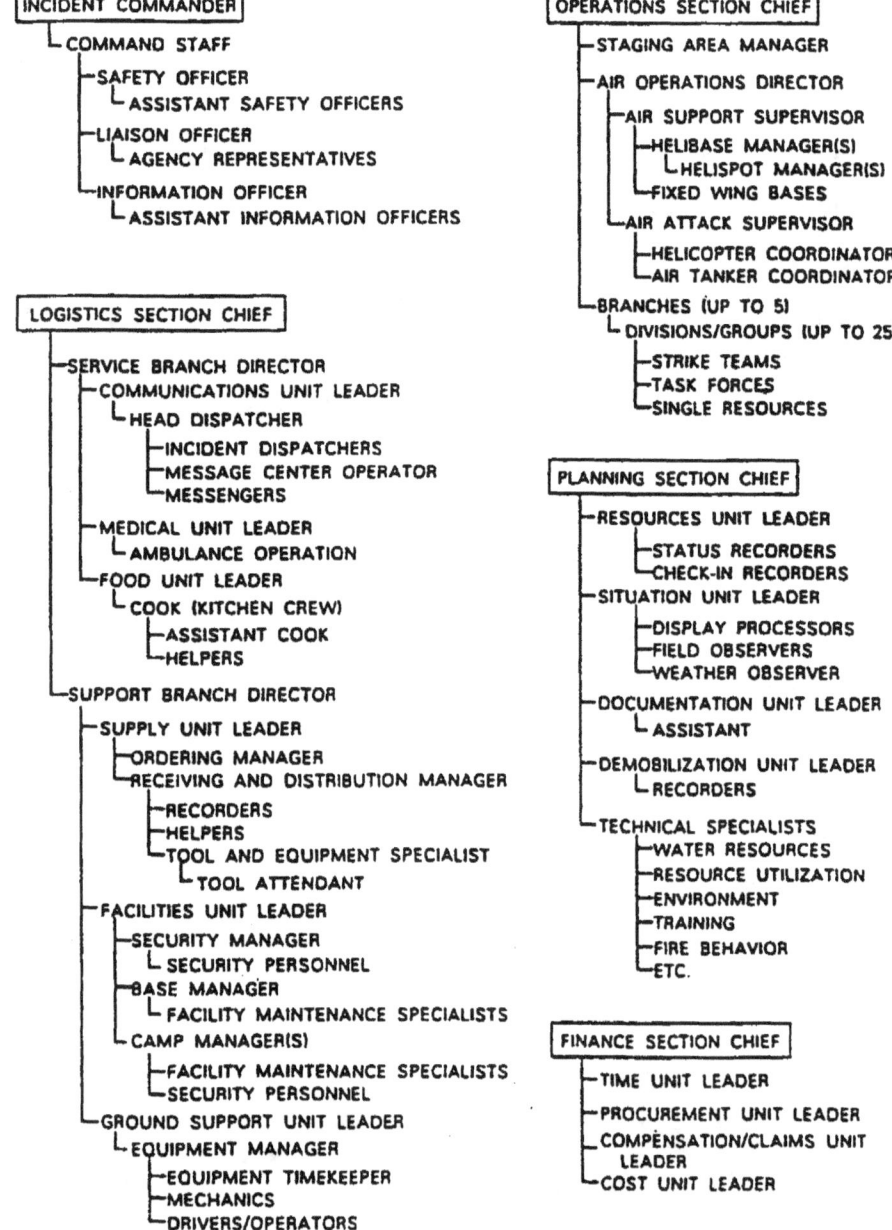

INCIDENT COMMANDER
└─COMMAND STAFF
  ├─SAFETY OFFICER
  │ └─ASSISTANT SAFETY OFFICERS
  ├─LIAISON OFFICER
  │ └─AGENCY REPRESENTATIVES
  └─INFORMATION OFFICER
     └─ASSISTANT INFORMATION OFFICERS

LOGISTICS SECTION CHIEF
├─SERVICE BRANCH DIRECTOR
│ ├─COMMUNICATIONS UNIT LEADER
│ │ └─HEAD DISPATCHER
│ │   ├─INCIDENT DISPATCHERS
│ │   ├─MESSAGE CENTER OPERATOR
│ │   └─MESSENGERS
│ ├─MEDICAL UNIT LEADER
│ │ └─AMBULANCE OPERATION
│ └─FOOD UNIT LEADER
│   └─COOK (KITCHEN CREW)
│   ├─ASSISTANT COOK
│   └─HELPERS
└─SUPPORT BRANCH DIRECTOR
   ├─SUPPLY UNIT LEADER
   │ ├─ORDERING MANAGER
   │ └─RECEIVING AND DISTRIBUTION MANAGER
   │   ├─RECORDERS
   │   ├─HELPERS
   │   └─TOOL AND EQUIPMENT SPECIALIST
   │   └─TOOL ATTENDANT
   ├─FACILITIES UNIT LEADER
   │ ├─SECURITY MANAGER
   │ │ └─SECURITY PERSONNEL
   │ ├─BASE MANAGER
   │ │ └─FACILITY MAINTENANCE SPECIALISTS
   │ └─CAMP MANAGER(S)
   │   ├─FACILITY MAINTENANCE SPECIALISTS
   │   └─SECURITY PERSONNEL
   └─GROUND SUPPORT UNIT LEADER
      └─EQUIPMENT MANAGER
         ├─EQUIPMENT TIMEKEEPER
         ├─MECHANICS
         └─DRIVERS/OPERATORS

OPERATIONS SECTION CHIEF
├─STAGING AREA MANAGER
├─AIR OPERATIONS DIRECTOR
│ ├─AIR SUPPORT SUPERVISOR
│ │ ├─HELIBASE MANAGER(S)
│ │ │ └─HELISPOT MANAGER(S)
│ │ └─FIXED WING BASES
│ └─AIR ATTACK SUPERVISOR
│   ├─HELICOPTER COORDINATOR
│   └─AIR TANKER COORDINATOR
└─BRANCHES (UP TO 5)
   └─DIVISIONS/GROUPS (UP TO 25)
      ├─STRIKE TEAMS
      ├─TASK FORCES
      └─SINGLE RESOURCES

PLANNING SECTION CHIEF
├─RESOURCES UNIT LEADER
│ ├─STATUS RECORDERS
│ └─CHECK-IN RECORDERS
├─SITUATION UNIT LEADER
│ ├─DISPLAY PROCESSORS
│ ├─FIELD OBSERVERS
│ └─WEATHER OBSERVER
├─DOCUMENTATION UNIT LEADER
│ └─ASSISTANT
├─DEMOBILIZATION UNIT LEADER
│ └─RECORDERS
└─TECHNICAL SPECIALISTS
   ├─WATER RESOURCES
   ├─RESOURCE UTILIZATION
   ├─ENVIRONMENT
   ├─TRAINING
   ├─FIRE BEHAVIOR
   └─ETC.

FINANCE SECTION CHIEF
├─TIME UNIT LEADER
├─PROCUREMENT UNIT LEADER
├─COMPENSATION/CLAIMS UNIT LEADER
└─COST UNIT LEADER

LINK BETWEEN ICS AND MUTUAL AID SYSTEM

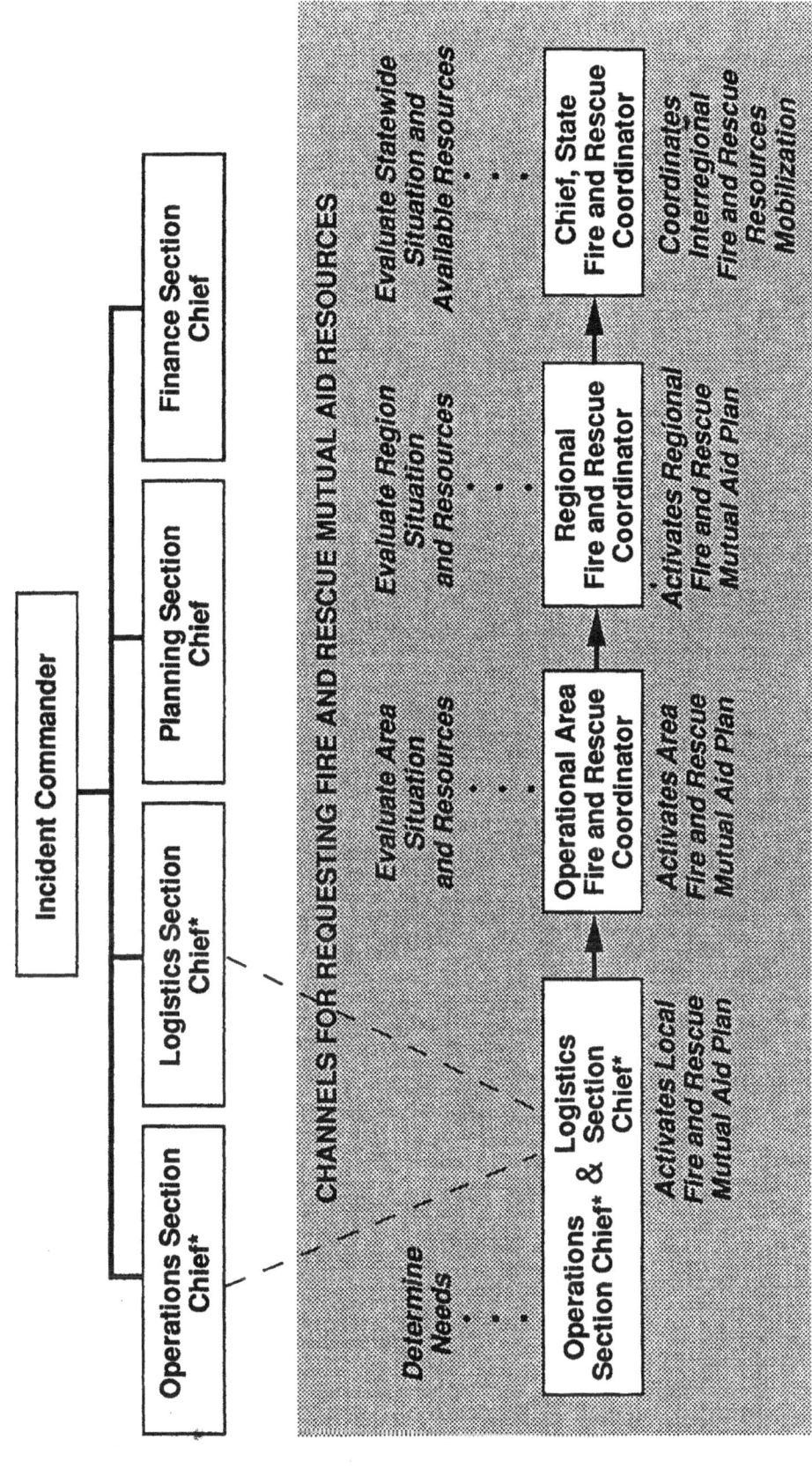

*The Operations Chief and the Logistics Chief serve as the links between the Incident Command System and the Fire and Rescue Mutual Aid System. In small, local incidents, one or both of these functions may be performed by the local Fire Chief or Incident Commander.

1879-5-13-91-9
R6-13-91

Appendix E

Map of Wildland Fire Protection Responsibility Areas

The State's Department of Forestry and Fire Protection protects more than 30 million acres of State-responsibility land in California and assists the Federal government on request to protect more than 44 million acres of Federal land within the State.

WILDLAND FIRE PROTECTION RESPONSIBILITY AREAS

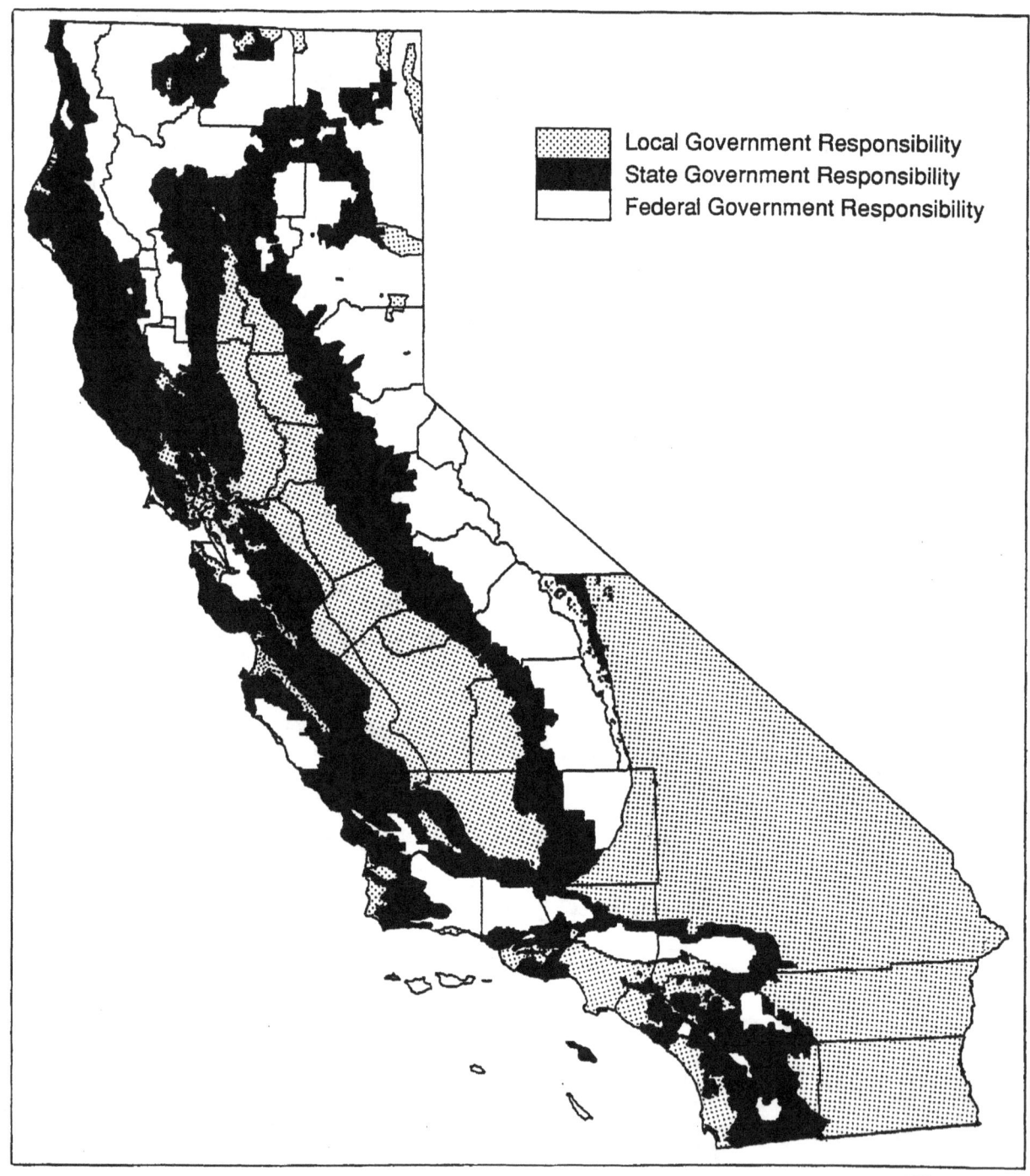

Local Government Responsibility
State Government Responsibility
Federal Government Responsibility

1879-5-13-91-11

Appendix F

"Paint" Fire Area Map

Map showing the "footprint" of the "Paint" fire in Los Padres National Forest/Santa Barbara County. (Reprinted with permission of the Santa Barbara News-Press.)

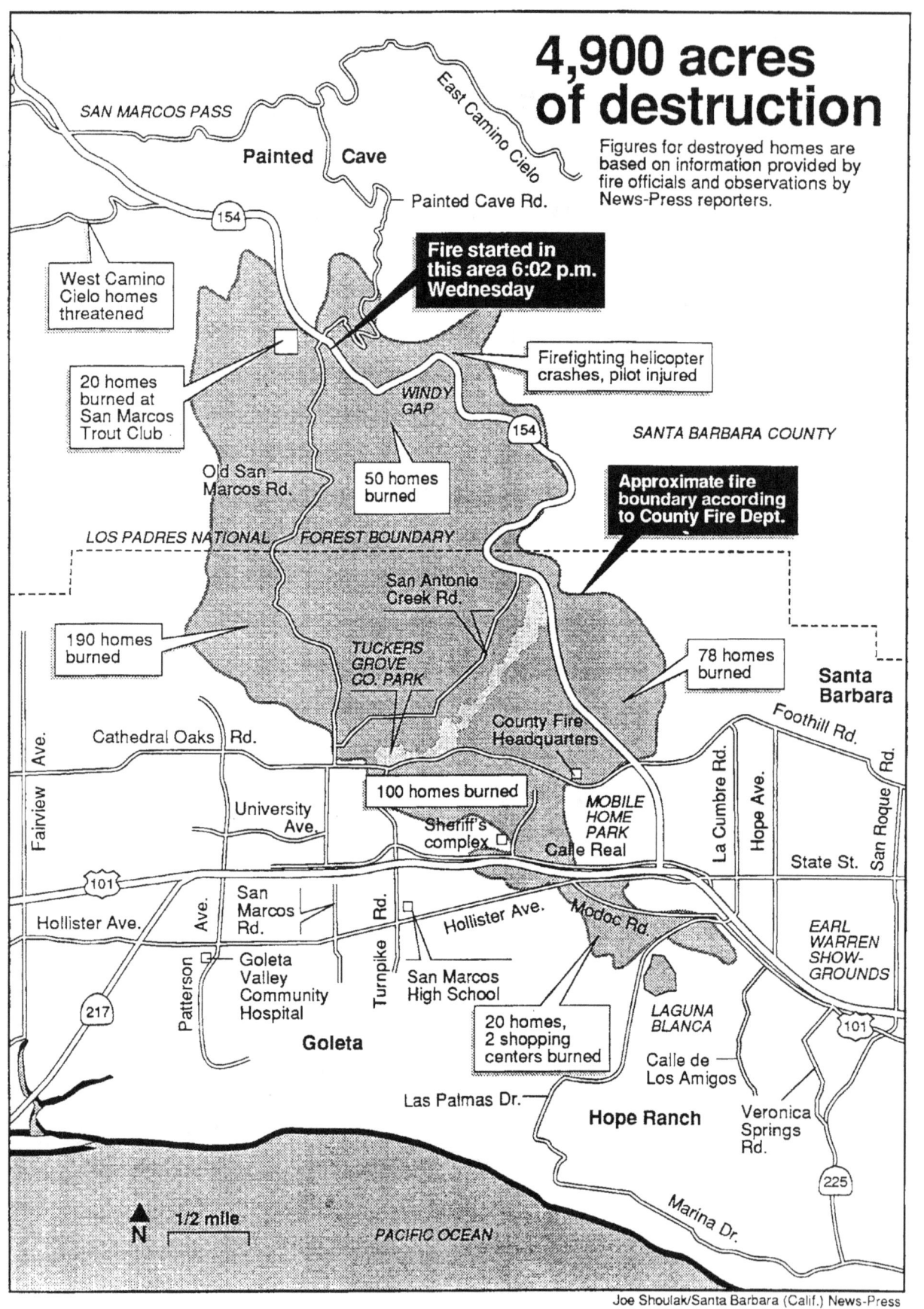

4,900 acres of destruction

Figures for destroyed homes are based on information provided by fire officials and observations by News-Press reporters.

SAN MARCOS PASS

East Camino Cielo

Painted   Cave

154

Painted Cave Rd.

Fire started in this area 6:02 p.m. Wednesday

West Camino Cielo homes threatened

Firefighting helicopter crashes, pilot injured

20 homes burned at San Marcos Trout Club

WINDY GAP

154

SANTA BARBARA COUNTY

Old San Marcos Rd.

50 homes burned

Approximate fire boundary according to County Fire Dept.

LOS PADRES NATIONAL   FOREST BOUNDARY

San Antonio Creek Rd.

190 homes burned

TUCKERS GROVE CO. PARK

78 homes burned

Santa Barbara

Foothill Rd.

Cathedral Oaks Rd.

County Fire Headquarters

Fairview Ave.

La Cumbre Rd.

Hope Ave.

San Roque Rd.

100 homes burned

MOBILE HOME PARK

Calle Real

University Ave.

Sheriff's complex

State St.

101

Hollister Ave.

San Marcos Rd.

Turnpike Rd.

Hollister Ave.

Modoc Rd.

EARL WARREN SHOW-GROUNDS

Patterson Ave.

Goleta Valley Community Hospital

San Marcos High School

LAGUNA BLANCA

101

217

Goleta

20 homes, 2 shopping centers burned

Calle de Los Amigos

Las Palmas Dr.

Hope Ranch

Veronica Springs Rd.

225

Marina Dr.

N   1/2 mile

PACIFIC OCEAN

Joe Shoulak/Santa Barbara (Calif.) News-Press

Appendix G

Loma Prieta Earthquake: Structural Collapse of I-880 and Status of Map of Trapped Vehicles

1. Twisted steel, broken concrete on one and one-half miles of I-880 presented a challenge for the major incident management team lead by the CDF that responded to a mutual aid request from the Oakland Fire Department following the 1989 Loma Prieta earthquake. (Photo courtesy of Janice Raymond, California Department of Transportation.)

2. An incident map, showing the position of vehicles trapped along the collapsed section of I-880 and the status of their occupants following the earthquake became an important tool for the incident management team. (Photo courtesy of Janice Raymond, CDF.)

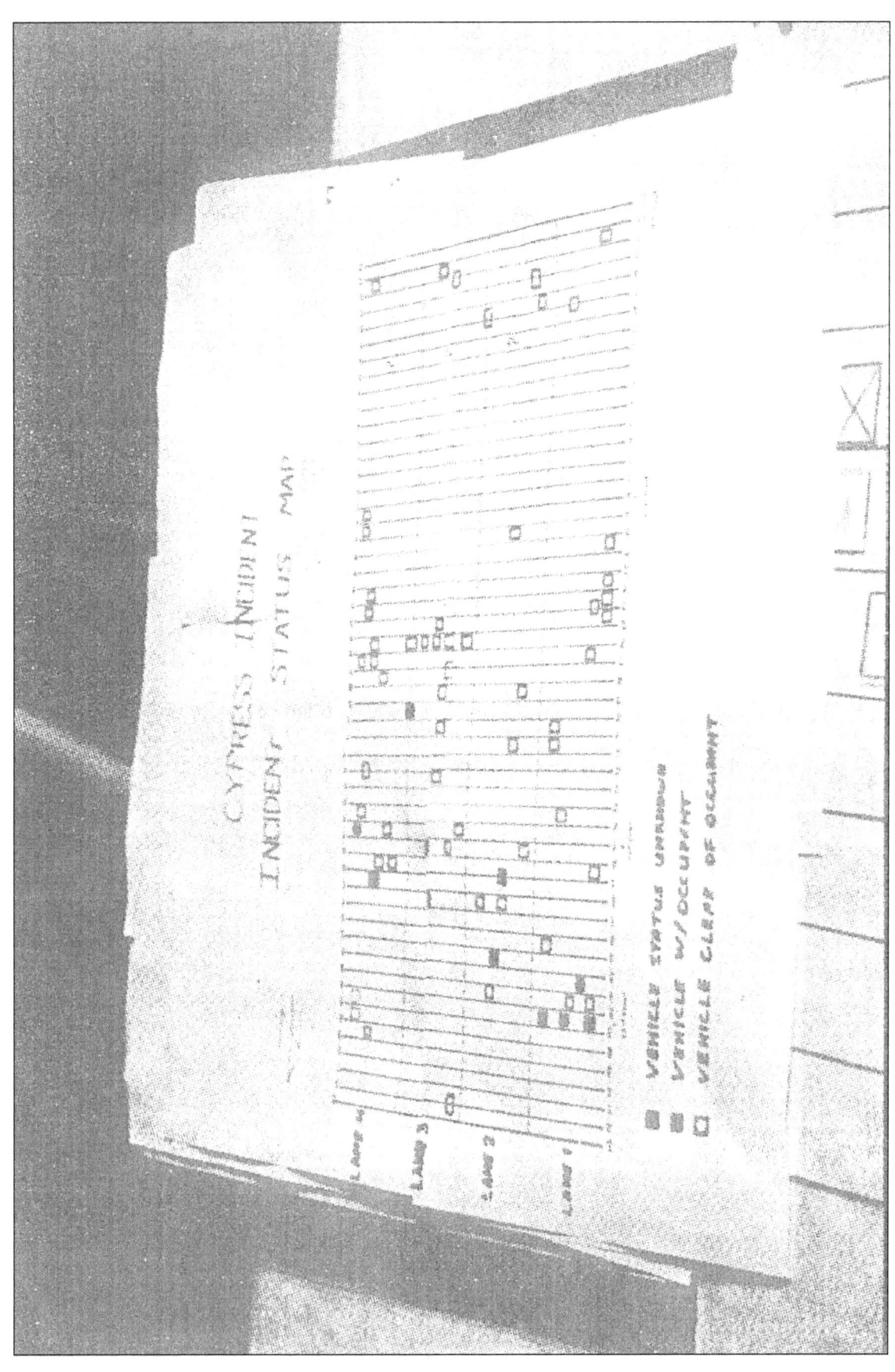

Appendix H

Bibliography

1. "CDF: Fire Protection...through cooperation," California Department of Forestry and Fire Protection, 1990

2. "Cooperative Agreement between State of California Office of Emergency Services; State of California Department of Forestry and Fire Protection, Pacific Southwest Region; USDA Forest Service; USDI Bureau of Land Management, California Office; and USDI National Park Service, Western Region," 1990

3. Cullom, Keith D., "On The Job – California," Firehouse, September 1990

4. "FIRESCOPE Decision Process and Operational Plan Manual: MACS 410-4," California Office of Emergency Services, 1990

5. "Hundreds Flee Wildfire, Arson launches 2-mile path of destruction," *Santa Barbara News-Press*, June 28, 1990

6. Perry, Don G., "Wildland Siege in Santa Barbara Co.," *American Fire Journal*, August 1990

7. Perry, Donald G., "Worst-Ever W/U Interface Fire in California Takes 648 Structures," *Wildfire News and Notes*, Vol. 4, No. 4, July-August 1990

8. Perez, Steven B., "California's Number 1 Fire Problem: Urban/Wildland Interface," *The California Fire Service*, October 1990

9. "The 1989 Loma Prieta Earthquake: The CDF Response," California Department of Forestry and Fire Protection, 199

10. "Election Process: Region and Operational Area Fire and Rescue Coordinators," State of California Office of Emergency Services Fire and Rescue Division, 1989

11. Garza, John, "Lessons Learned from the Quake – Interviews with some of the Chiefs," *American Fire Journal*, December 1989

12. Garza, John V., "Surrounding Communities Hit Hard by the Quake," *American Fire Journal*, December 1989

13. "Incident Command System Field Operations Guide: ICS 420-1" 1989

14. "Orange County Fire Services Operational Area Mutual Aid Plan," Orange County Fire Department, 1989

15. "Resource Designation System, Statewide Fire and Rescue Mutual Aid System: MACS 410-2," California Office of Emergency Services, 198

16. Sherman, Edward R., "A B/C's Story," *American Fire Journal*, December 1989

17. Sparks, Carl A., "Automatic Aid: An Approach to Improved Fire Resources," Research paper for Strategic Analysis of Fire Department Operations course, National Fire Academy, 1989

18. U.S. Fire Administration, Office of Policy and Coordination, "The Loma Prieta Quake: Emergency Response and Stabilization," *American Fire Journal*, December 1989

19. "California Fire Service and Rescue Emergency Mutual Aid Plan," California Office of Emergency Services, 1988

20. "Fire and Rescue Mutual Aid System," California Office of Emergency Services, 1988

21. "FIRESCOPE Past, Current and Future Directions: A Progress Report," California Office of Emergency Services, 1988

22. "History and Organization," State of California Office of Emergency Services Fire and Rescue Division, 1982, Revised 1988

23. "Proposed 5-Year FIRESCOPE Expansion and Implementation Plan," California Office of Emergency Services, 1988

24. "Strike Team Leader, Statewide Fire and Rescue Mutual Aid System," California Office of Emergency Services, 1988

25. "Final Report: Needs Assessment and Study of the FIRESCOPE Program," Ryland Research, Inc., 1987

26. "Exemplary Practices in Emergency Management: The California FIRESCOPE Program," Emergency Management Institute, Federal Emergency Management Agency, Monograph Series No. 1, 1986

27. "MACS Procedures Guide: MACS 410-1," California Office of Emergency Services, 1986

28. "California Fire and Rescue Emergency Plan Mutual Aid Interface with FIRESCOPE," California Office of Emergency Services, 1984

29. "Red Flag Alert Notification Procedures Guide: MACS 410-3," California Office of Emergency Services, 1984

30. Snell, Larry and Associates, "An Overview and Assessment of the California Master Mutual Aid Program," A report to the California Seismic Safety Commission, 1979

www.ingramcontent.com/pod-product-compliance
Lightning Source LLC
Chambersburg PA
CBHW081619170526
45166CB00009B/3026